YOUR KNOWLEDGE HAS VALUE

- We will publish your bachelor's and master's thesis, essays and papers

- Your own eBook and book - sold worldwide in all relevant shops

- Earn money with each sale

Upload your text at www.GRIN.com and publish for free

Bibliographic information published by the German National Library:

The German National Library lists this publication in the National Bibliography;
detailed bibliographic data are available on the Internet at http://dnb.dnb.de .

Imprint:

Copyright © 2011 GRIN Verlag, Open Publishing GmbH
Print and binding: Books on Demand GmbH, Norderstedt Germany
ISBN: 978-3-668-10671-0

This book at GRIN:

http://www.grin.com/en/e-book/193180/assessment-of-biomass-for-energy-purposes-
for-jawor-county

Lukasz Niedzwiecki

Assessment of biomass for energy purposes for Jawor county

Availability of biomass for energy resources and feasibility to transfer Jawor from fossil fuels to biofuels

GRIN Publishing

Linnæus University

4BT001 Assessment of biomass resources

Availability of biomass for energy resources and feasibility to transfer Jawor from fossil fuels to biofuels

Lukasz Niedzwiecki

Table of contents

1. Introduction

1.1 Administrative division of Poland

The administrative division of Poland since 1999 has been based on three levels of subdivision. The territory of Poland is divided into voivodeships (regions); these are further divided into powiats (counties), and these in turn are divided into gminas (communes or municipalities). Major cities normally have the status of both gmina and powiat. Poland currently has 16 voivodeships, 379 powiats (including 65 cities with powiat status), and 2 478 gminas.

1.2 Jawor

Aim of this essay is to estimate availability of biomass for energy purposes in my hometown Jawor. Jawor is a small Town inhabited by 24 576 of citizens which is 8 724 households [6]. Jawor is located in south-western part of Poland in region called Dolny Śląsk (eng. Lower Silesia) - Jawor doesn't have powiat status itself but is a capital of Jawor county consisting of 6 communes. Jawor itself has a status of municipality (municipal commune)(fig 1.1). County in total is inhabited approximately with 53 000 people.

Figure 1.1 Map of Lower Silesia, with location of Jawor county and all communes belonging to the county

1.3 Methodology

Some resources are estimated only in terms of the county, other are estimated in terms of the region. Division has been made depending on biomass type, it's suitability for transport and local transport conditions. The distance from furthest place in region to Jawor does not exceed 150 km and in average could be considered less than 100 km. The distance from furthest place in county to Jawor does not exceed 25 km and in average could be considered 10 km.

2. Usefull energy and energy services in Jawor

2.1 General situation

Jawor is a typical example of a small town in south-western Poland. All households as well as all public facilities like schools, hospital etc. along with all the enterprises within the city area are connected to electricity grid. District heating covers more than half of the town. Most of the town is connected to national natural gas distribution system.

I assume covering only energy for households public facilities and enterprises, energy for transport is neglected. In Poland summer is not extremely hot, temperatures exceeding 30 °C happen sometimes but not often. In Jawor there are no big companies with big office buildings, there are also no big food industry facilities and no big warehouses dealing storing food that could possibly consume cooling. Climate cooling because of investment cost is not very popular, so i assume electricity and natural gas for climate cooling as 0.

2.2 Electricity

According to [9] mean usage of electricity by households in Poland is 1624 kWh/year. For 8 724 households it brings little bit less than **14,2 GWh/year** of electricity for Jawor. As mentioned before i assume electricity usage for climate cooling as 0. Prices of electricity in Poland are, comparing to salaries, relatively high so electric heating is not very popular. Most of the household electricity in typical polish small town household is consumed by household appliances such us fridges, washing machines etc. as well as home electronics f. ex. TV. Electric stoves are quite common but usually when household is connected to natural gas system gas stove is used for cooking. Usually people have both gas stove and gas oven, but sometimes "hybrids" such as gas stove and electric oven happen (my home might be an example).

My assumption is:
- electricity to produce hot tap water $\qquad$ 10 % of the total electricity $Q_{EL\,1}$
- electricity for comfort/climate cooling $\qquad$ 0 % of the total electricity $Q_{EL\,2}$
- electricity for comfort/climate heating $\qquad$ 5 % of the total electricity $Q_{EL\,3}$
- electricity for cooking $\qquad$ 5 % of the total electricity $Q_{EL\,4}$
- electricity for household appliances, electronics and light $\qquad$ 80 % of the total electricity $Q_{EL\,5}$

2.3 Natural gas

In Jawor most of the households and small entrepreneurs are connected to natural gas network system. There are also few big industrial users Kuźnia Jawor S.A., Korpo S.A. and IS Polska S.A. There is no data about their natural gas usage available. Kuźnia Jawor S.A. is very close to bankruptcy because of recent economical crisis and it would be very difficult to estimate their natural gas usage because it depends on production rate.

Table 2.1 - Natural gas usage in Jawor[6]

	Natural Gas per household per year	
2002	525,1	[m^3]
2003	455,3	[m^3]
2004	435,9	[m^3]
2005	430,2	[m^3]
2006	406,3	[m^3]
2007	377,8	[m^3]
2008	353,7	[m^3]
2009	399,8	[m^3]

mean value	423,0	[m^3]
standard deviation	52,6	[m^3]
safe value	633,3	[m^3]
Households	8 724	
Heating Value	39,5	[MJ/m^3]
Energy:	218 237,65	GJ / year
	60,62	GWh / year

High amounts of natural are used for heating in households not connected to district heating there are high fluctuations between different years in terms of yearly gas usage depending on temperatures during late autumn and winter. To estimate probable natural gas demand i used Gaussian distribution. I used $mean\ value + 4 \cdot standard\ deviation$ as a "safe value" and in that case, in my opinion, it's highly unlikely that yearly demand will be higher in practice. Natural gas is also used to produce hot tap water in some household that are supplied with heat for comfort heating purposes by district heating (my home might be an example).

My assumption is:
- gas to produce hot tap water　　　　　　　$40\ \%\ of\ the\ total\ gas\ \ Q_{GAS\,1}$
- gas for comfort/climate cooling　　　　　　$0\ \%\ of\ the\ total\ gas\ \ Q_{GAS\,2}$
- gas for comfort/climate heating　　　　　　$40\ \%\ of\ the\ total\ gas\ Q_{GAS\,3}$
- gas for cooking　　　　　　　　　　　　　$20\ \%\ of\ the\ total\ gas\ Q_{GAS\,4}$

2.4 Heating

District heating is supplied with heat by two heating plants. First is equipped with four grate boilers: 2 x WR5 and 2 x WR10. Second is equipped with four boilers: 2 x WR5 and 2x WR2.5. These types of boilers produce only hot water, not steam therefore plant produces only heat. Coal is used as a fuel.

Nominal power of boilers:
- WR 2.5　　-　　$2,9\ MW_{th}$
- WR 5　　　-　　$5,8\ MW_{th}$
- WR 10　　-　　$11,6\ MW_{th}$

Both heating plants are "oversized". They were built during communism times ca. 30 years ago. Those days system pipelines were not as good insulated. Insulation of buildings was not as good as it is nowadays. Those days there were also plans to build more blocks of flats in Jawor and to expand the city.

Nowadays both plants are connected with a big pipeline and they work alternately. Bigger plant works in so called "winter mode" - during heating season supplies heat used for district heating and hot tap water in households and public facilities. It also supplies heat for a swimming pool. When heating season is over smaller starts operation in so called "summer mode" and supplies heating for hot tap water and swimming pool (swimming pool facility uses less heat in warmer periods of year). Sometimes it may also deliver heat for households and public facilities, if the outside temperature gets too low during early autumn or late spring, but it does not happen often and does not require as much heat as it is during winter time.

In those parts of the city that are not connected to the district heating there are few boilers run by administrators of the objects that they deliver heat into, and these are: local housing co-operatives, town hospital, few of the town schools and kindergartens and other public facilities. Some households especially freestanding houses have their own source of heat. Fuel varies for all those cases but usually coal and natural gas seems to be dominant.

Table 2.2 - Heat production in Jawor county [6]

Teritorial unit	From own sources		From other sources	
	1999	2000	1999	2000
	[GJ]	[GJ]	[GJ]	[GJ]
Jawor county	176 416,0	150 102,0	140 477,0	137 701,0

Table 2.3 - Example data showing heat production in bigger heat plant during "winter mode"

February 2010	Mean daily temperature (outside)	Heat production	Heating value of coal
Date:	[℃]	[GJ]	[kJ/kg]
2010-02-01	-0,2	1313,6	21134
2010-02-02	-0,5	1328,1	21134
2010-02-03	0,7	1249,6	21134
2010-02-04	1,7	1272,4	21134
2010-02-05	2,6	1212,8	21134
2010-02-06	1,0	1320,3	21134
2010-02-07	0,0	1408,9	21134
2010-02-08	-5,9	1365,0	21134
2010-02-09	-5,2	1497,6	21134
2010-02-10	-2,4	1319,6	21134
2010-02-11	-4,6	1391,1	21134
2010-02-12	-1,6	1346,9	21134
2010-02-13	-1,3	1286,2	21134
2010-02-14	-1,5	1279,4	21134
2010-02-15	-0,3	1239,7	21134
2010-02-16	-0,7	1241,0	21134
2010-02-17	1,4	1211,8	21134
2010-02-18	3,2	1089,5	21134
2010-02-19	4,8	1036,5	21134
2010-02-20	2,8	1088,9	21134
2010-02-21	2,1	1093,1	21134

Data about the heating in GUS database (Tab2.2) [6] exist only for county as a whole and only for years 1999 - 2000 (census) data relates only to households and does not include industrial users, swimming pool and other public facilities. Data from district heating (Tab 2.3) do not give full picture since not all individual users are connected into. Data in Tab 2.2 are divided as own sources, which means that heat consumer produced the heat on his own, and other sources which mean that consumer is not the owner of the source which not necessarily mean that consumer was connected to district heating. Supply "from other sources" does not necessarily mean that customer was connected to the district heating - it might also mean that source belongs to local housing co-operative that household is a part of.

My estimation is: **380 000 GJ/year** that is **105,5 GWh/year** - that may be considered as overestimation but since supply is mandatory to meet the demands that overestimation seems to be justified. Nobody wants to be in a situation that he cannot receive energy service he needs.

My assumption is:
- gas to produce hot tap water $40\% \ of \ the \ total \ gas \ Q_{DH\,1}$
- gas for comfort/climate heating $60\% \ of \ the \ total \ gas \ Q_{DH\,3}$

2.5 Energy service demand

Table 2.4 - Energy service demand - summary

	%	GWh / year	GJ/ year
Q el total	100	14,20	51120,0
Q el1	10	1,42	5112,0
Q el2	0	0,00	0,0
Q el3	5	0,71	2556,0
Q el4	5	0,71	2556,0
Q el5	80	11,36	40896,0
Q gas total	100	60,62	218232,0
Q gas1	40	24,25	87292,8
Q gas2	0	0	0,0
Q gas3	40	24,25	87292,8
Q gas4	20	12,12	43646,4
Q dh total	100	105,50	380000,0
Q dh1	40	42,20	152000,0
Q dh3	60	63,30	228000,0

In Tab 2.4 there are summarised different kinds of energy services. Interesting thing is that some of these services we could be delivered using the same type of useful energy. These types are grouped in Tab 2.5. Q_{th} is low temperature heat and $Q_{hi\ th}$ is high temperature heat (as high as usual stove/oven temperature).

Table 2.5 - Useful energy demand - summary

	GWh / year	GJ / year
Q th	156,1	562 053,6
Q el	11,4	40 896,0
Q hi th	12,8	46 202,4

3. Biomass resources available in Lower Silesia region

3.1 Agricultural biomass

Table 3.1 - Crops in Lower Silesia [6] and straw energy potential per year

	wheat	rye	barley	oat	triticale	mixed crops
	[t]	[t]	[t]	[t]	[t]	[t]
1999	1 186 468,2	178 494,9	286 787,8	58 209,6	67 231,3	98 976,0
2000	1 172 530,5	146 961,2	257 875,2	54 505,2	57 489,7	81 929,2
2001	1 208 889,0	146 884,5	260 579,4	49 568,4	59 714,5	80 023,5
2002	1 236 756,9	114 877,5	235 603,7	67 856,8	62 284,0	78 440,7
2003	1 034 580,9	105 575,0	211 329,4	54 410,5	52 537,9	70 211,3
2004	1 376 095,2	146 545,9	293 082,0	68 288,0	74 780,8	90 108,4
2005	1 211 119,4	133 609,0	360 274,7	77 108,6	98 444,0	92 611,4
2006	924 441,2	93 078,7	258 401,8	53 787,7	74 016,5	69 880,8
2007	1 124 575,8	134 937,5	298 013,3	64 433,8	88 232,6	74 236,7
2008	1 170 332,6	135 714,2	304 198,6	58 896,2	88 476,5	59 395,6
2009	1 261 474,1	145 724,8	298 587,2	70 517,4	105 961,4	69 162,0

mean value	1 173 387,6	134 763,9	278 612,1	61 598,4	75 379,0	78 634,1
standard deviation	118 048,6	23 354,7	39 894,1	8 594,0	17 695,3	11 749,0
safe value [t]	937 290,5	88 054,5	198 823,8	44 410,4	39 988,4	55 136,2
RPR	0,81	1,24	0,67	1,01	0,94	0,67
straw [t]	759 205,27	109 187,59	133 211,97	44 854,52	37 589,12	36 941,2
straw available [t]	227 761,58	32 756,28	39 963,59	13 456,35	11 276,74	11 082,3
Energy:						
TJ	11 008,48	1 583,22	1 931,57	650,39	545,04	535,65
GWh	3 057,91	439,78	536,55	180,66	151,40	148,79

	TJ / year	GWh / year
Total	16 254,35	4 515,10

In Poland crops are being sown in two different periods thus are divided into two groups: "ozime" and "jare". Crops called "ozime" are sown in late autumn and harvested in early summer. Crops called "jare" are sown in spring and harvested late summer and early autumn. In yield database [6] crops are grouped by species so "ozime" and "jare" yield of the same species is added one to another. According to [4] there is a difference between Residue to Product Ratio between "ozime" and "jare" crops among the same species. RPR also varies depending on yield per ha among the same species [4]. Value of RPR used for calculation was always the lowest one for every species.

As safe value i consider **_mean value − 2 · standard deviation_** which gives a reasonably good probability of achieving desired straw yield according to Gaussian distribution. Standard deviation was calculated using Standard Deviation function of MS Excel.

Straw in Poland is partially used in agriculture for animal feeding and bedding and also as a fertilizer. Cautious estimate [4] says that 30% of straw could be probably used for energy production - that is Straw Available position in Table 3.1.

Quality of straw varies depending on weather conditions during harvest time, storage and many other factors. Yield of the crops from GUS database [6] and RPR values refer to biomass harvested from the field i assumed Lower Heating Value of 14,5 GJ/tonne as a safe value. Biomass would definitely be dried. Most probably it would be baled and left for some time on sight to dry in open-air drying process. It's a commonly practiced, no matter if straw is planned to be utilised as a fuel or for other purposes. Due to open air drying biomass would lose some amount of water and heating value is chosen to compensate that effect - disadvantageous situation in terms of moisture is assumed.

3.2 Woody biomass

In Poland most of the forest area belongs to the state and is being managed by State Forests Agency, some are property of other institutions, communes and municipalities and private owners - Fig.3.1. Each Regional Division of State Forests Agency is responsible for it's own region and manages it according to Regional operational plan for state silvicultural policy [2] which is based on National Forestation Programme [8]. Each Regional division consist of subdivisions called "nadleśnictwa" - these units are not divided in the exactly same way as counties. Each of those subdivision runs according to document called "Instrukcja urządzania lasu" (eng. Instruction of forest management). Forestation of Poland is 29,0% of country's area and according to [8] should reach 30% in 2020 and 33% in 2050.

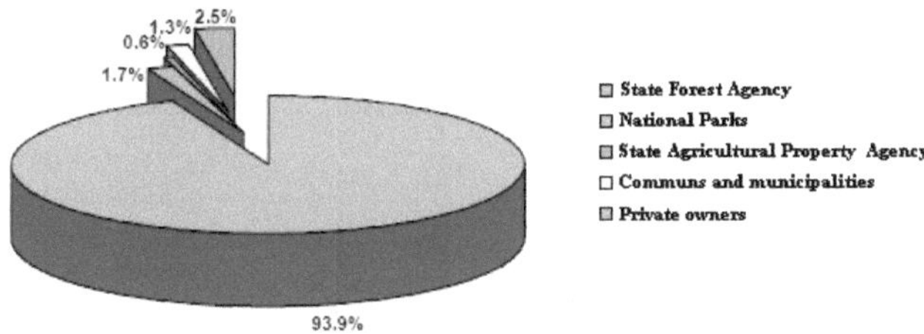

Silvicultural Policy is made according to two main rules [2]:
- balance and sustainability
- multifunctional role of forest

Forestry operations are usually thinning and prunings. Clear cuts are used usually to make space for some new infrastructure that has to go through the forest - ex. high voltage lines, natural gas pipelines or roads. Clear cuts are also used to make clear spaces between parts of forest to stop eventual fires from spreading.

One of the main uncertainties is amount of wood cut because of forest health reasons. During communism, especially soon after war most of the newly forested area consisted of monocultures - mainly pine. Monocultures are less resistant to pest attacks like insects or fungi. Amount of wood cut because of forest protection may vary depending on pest population, but total amount in future would most likely decrease, because biodiversity is an important part of new regulations [2, 8].

According to [4] there are 5 groups of woody biomass:
- *Forestry operations residues* - these are residues that are left after forestry operations like thinning, prunning and clear cutting. Residues are produced during felling and logging. There is no official record about usage of that group, according to [4] except collection of the residues by nearby inhabitants there is not much interest in that resource but in some areas that type of collection might be significant. As a safe value [4] assumes collection of 50% of possible resource.
- *Industrial operations residues* - there are basically no pulp and paper companies in Lower Silesia so these are mostly sawmills residues. It's assumed [4] that 87% of residues is used by sawmills.
- *Recycled wood* - this group consist of demolition wood, broken furniture, broken pallets etc... It is assumed [4] that 30% is not used and not contaminated.
- *Orchard operations residues* - these are residues that come from agricultural operations in orchards.
- *Roadside operation residues* - in Poland there are trees usually grown between the roadside and the fields outside inhabited areas.

Table 3.2 - woody biomass available for energy purposes [4]

	Jawor county		Lower Silesia	
	TJ	GWh	TJ	GWh
Forestry op. residues	6,0	1,67	276,0	76,7
Industrial residues	2,0	0,56	130,0	36,1
Recycled wood	12,0	3,33	661,0	183,6
Orchard op. residues	0,1	0,03	5,4	1,5
Roadside op. residues	4,0	1,11	108,9	30,3
TOTAL	**24,1**	**6,69**	**1 181,3**	**328,1**

Lower Heating Value of $15\ ^{GJ}/_{tonne}$ and bulk density of fresh biomass $250\ kg/m^3$ was assumed [4].

Part of the wood cut by State Forests Agency is being sold as a fuel. This is type S4. According to Polish technical standards S4 is a wood that has diameter between 5 and 24 cm without the bark. It has been sold only for household use. It has usually been bought by people that owe fireplaces and wood boilers both old and new type. It has not been considered as biomass in terms of energy producers. Recently in Poland there were debates about new regulations that could change that situation and allow energy producers to use every type of wood. Wood called "Drewno opałowe" [4] was neglected in this assessment, although in my opinion uncertainty of future legislation is a significant constrain in terms of using woody biomass as a fuel in Poland.

4. Biomass resources available in Jawor county

4.1 Dung

Dung is different than other types of biomass with agricultural origin. It has extremely high moisture and ash content and represents negative heating value as such, although using proper technology discussed further it might be utilized as an energy source.

Table 4.1 - Number of farmers keeping animals in the county and in the region [6]; county to region ratio

Teritorial unit	Farmers keeping animals			
	cattle	pigs	horses	sheep
	2002	2002	2002	2002
	[-]	[-]	[-]	[-]
Lower Silesia region	21 773	24 854	3 791	757
Jawor	10	16	0	0
Bolków	235	177	54	16
Męcinka	203	219	16	6
Mściwojów	126	190	5	0
Paszowice	134	138	12	5
Wądroże Wielkie	156	201	0	0
Jawor county	864	941	87	27
county/region [%]	3,97	3,79	2,29	3,57

There is no data available of animal stock in the county. There are data about animal stock in the region and there is number of farmers keeping animals available both for region and county [6]. I estimated county to region ratio of farmers keeping animals. Assuming that number of farmers keeping animals is proportional to number of animals kept i estimated amount of animals in the county.

To estimate amount of dung produced by each animal i used amounts of straw used for bedding [4]. I assumed ash content of 50%, moisture of 90% and energy content of dung as 19,5 GJ/t for dry ash free substance.

Table 4.2 - Dung potential in the county based on animal stock in the region [6] and county/region ratio

	Cattle	Pigs	Sheep	Goats	Poultry	Horses
1999	173 038	580 590	15 415	-	4 669 075	12 048
2000	151 533	472 183	14 667	-	4 787 751	12 048
2001	137 810	542 702	11 178	-	5 071 720	12 353
2002	126 052	548 279	13 490	-	6 669 686	9 408
2003	119 210	523 035	12 722	-	6 062 920	9 536
2004	118 118	484 527	10 525	11 475	5 290 606	8 506
2005	114 631	457 102	8 357	7 571	6 434 721	11 393
2006	106 138	437 783	8 909	8 012	5 192 009	11 703
2007	110 155	416 829	11 951	8 642	5 796 484	13 054
2008	105 900	307 768	10 218	7 765	5 977 623	13 150
2009	98 932	326 421	9 373	7 729	5 336 928	12 494
mean value	123774,27	463383,55	11527,73	8532,33	5571774,82	11426,64
standard deviation	22291,76	87838,42	2341,35	1489,99	660648,30	1569,87
safe value	56898,98	199868,27	4503,68	4062,36	3589829,92	6717,03
county/region [%]	3,97	3,79	3,57	3,57	3,00	2,29
units in county	2257,88	7567,23	160,63	144,89	107694,90	154,15
dung per unit [t/year]	2,55	0,93	0,45	0,45	0,02	4,00
dung [t/year]	5757,58	7037,53	72,28	65,20	2153,90	616,60
DAF tonnes/year	287,88	351,88	3,61	3,26	107,69	30,83
GJ	5613,64	6861,59	70,48	63,57	2100,05	601,18
GWh	1,56	1,91	0,02	0,02	0,58	0,17
TOTAL:	15310,52	GJ / year				
	4,25	GWh / year				

There are many factors that animal stock is dependent on. Farmers make their decisions about animal stock they keep upon a prices of meat and other goods produced by animals like for example milk or wool. Also regulation policy of EU and Polish government is important.

As "safe value" i consider $mean\ value - 3 \cdot standard\ deviation$ which gives a reasonably good probability of achieving desired straw yield according to Gaussian distribution. Standard deviation was calculated using Standard Deviation function of MS Excel. Due to incomplete data mentioned before estimation is very rough. Although overall trend of animal stock is decreasing it will probably stop at some point since smaller stock overall would most likely lead to higher prices of animal products and change the tendency. Also "safe values" are still much lower than smallest values noted so it still seems to be underestimated at that point.

4.2 Sewage sludge

There are 7 sewage treatment plants in the county [1]. At present sewage sludge is used for recultivation of poor quality soil within the county area [1] in approximately 97%.

Sewage sludge potential in the county [1]:
- o 1 471 tons of dry substance in 2004
- o 1 205 tons of dry substance in 2005
- o 1 458 tons of dry substance in 2006

Assuming sewage sludge yield at safe value of 1 200 tons DAF/year, ash content 50% and heating value for dry ash free substance 19,5 GJ/t energy potential of sewage sludge in Jawor county may be considered:

$$11\ 700\ ^{GJ}/_{year} \quad \text{which is} \quad 3,25\ ^{GWh}/_{year}$$

4.3 Municipal Solid Waste

In Jawor county there are 10 landfills able to store waste classified as non hazardous waste [1]. Four are being used, four are already closed, one is ready but not used at the moment and one is being recultivated - Fig 4.1.

Figure 4.1 - Landfills in Jawor county

1.	Składowisko Odpadów Komunalnych dla miasta Jawora ul. Słowackiego, 59-400 Jawor	Jawor	operating
2.	Składowisko Odpadów Komunalnych w Wierzchosławicach	Bolków	operating
3.	Składowisko Muchów	Męcinka	closed
4.	Składowisko Sichów	Męcinka	closed
5.	Składowisko Drzymałowice	Mściwojów	closed
6.	Składowisko Paszowice	Paszowice	closed
7.	Składowisko Nowa Wieś Wielka	Paszowice	new
8.	Składowisko Wądroże Małe	Wądroże Wielkie	operating
9.	Składowisko Budziszów Wielki	Wądroże Wielkie	operating
10.	Składowisko Mierczyce	Wądroże Wielkie	closed, recultivated

Jawor county plan of waste management [1] shows some potential in that area. It also shows potential synergy. Biodegradable waste are divided into three groups:
- green wastes from gardens and parks (which is small size woody biomass and leafs fallen in the autumn along with hay leftovers after mowing municipal green terrains - mostly parks)
- paper and paperboard (usually residues from supermarkets)
- organic waste from co called "mixed stream" (which consist mainly of household organic wastes, multimaterial wastes and textiles)

Paper and paperboard residues could possibly be used as an energy source, but i neglect them because it is possible to recycle and reuse them, although it's not being done at 100% of them at the moment.

Table 4.3 - Organic waste from co called "mixed stream"- prognosis [1]

Organic waste		
2007	5 828,0	tonnes / year
2008	5 847,0	tonnes / year
2009	5 859,0	tonnes / year
2010	5 871,0	tonnes / year
2011	5 882,0	tonnes / year
2012	5 937,0	tonnes / year
2013	5 992,0	tonnes / year
2014	6 046,0	tonnes / year
2015	6 101,0	tonnes / year
mean value	5 929,2	tonnes / year
standard deviation	96,6	tonnes / year
safe value	5 639,3	tonnes / year
Heating value	5,0	GJ/t
Total potential:	**28 196,6**	**GJ / year**
	7,83	**GWh / year**

As "safe value" i consider ***mean value − 3 · standard deviation*** which gives a reasonably good probability of achieving desired straw yield according to Gaussian distribution. Standard deviation was calculated using Standard Deviation function of MS Excel. It seems to be underestimated especially considering trend that, according to the prognosis [1], amount of waste in that stream is going to increase. There is quite low heating value 5 GJ/t assumed [10].

Table 4.4 - Prognosis for waste and waste management plan to meet necessary requirements [1]

Category	Amount of biodegradable wastes [tonnes]									
	1995	2007	2008	2009	2010	2011	2012	2013	2014	2015
Total amount of biodegradable waste produced in the county	5 989,1	8 908,8	8 939,1	8 957,3	8 974,8	8 991,6	9 075,6	9 159,5	9 243,5	9 327,5
Amount of biodegradable wastes possible to be placed on landfills according to regulations	-	5 390,2	5 090,7	4 791,3	4 491,8	3 892,9	3 294,0	2 994,6	2 695,1	2 395,6
Amount of paper recycled	-	953,3	976,4	998,3	1 040.2	1 082,2	1 132,8	1 184,1	1 236,2	1 247,5
Amount of surplus waste that needs to be recovered to meet the regulations	-	2 565,3	2 872,0	3 167,7	3 442,7	4 016,5	4 648,8	4 980,8	5 312,2	5 684,4
Total paper and paperboard	-	3 072	3 083	3 089	3 095	3 101	3 130	3 159	3 188	3 217

Prognosis in Tab 4.4 shows that even if all paper is going to be recycled it would be hard to meet upper limits for landfill storage set by Waste management plan and requirements set by waste management regulations. Therefore there is a possibility of synergy between energy production and waste management as a whole. That situation is typical for whole Poland at present.

Possibility of using that resource relies much on proper waste collection and sorting system. It would require additional investment and some kind of environmental informational campaign to increase the awareness among the local society, but more environmental gains than just clean energy could possibly be achieved that way.

5. Technologies suitable to supply energy services in terms of biomass availability

5.1 Meeting demands with the potential

Table 5.1 - Comparison between potential and demands

POTENTIAL				DEMANDS		
		TJ / year	GWh / year		GWh / year	GJ / year
Lower Silesia	Straw	16 254,35	4 515,10	Q th	156,1	562 053,60
	Wood	1 181,30	328,1	Q el	11,4	40 896,00
				Q hi th	12,8	46 202,40
		GJ / year	GWh / year			
Jawor county	Dung	15 310,52	4,25			
	Sewage sludge	11 700,00	3,25			
	Municipal Solid Waste	28 196,60	7,83			

First look at the Tab. 5.1 would bring the conclusion that it's possible to switch from fossil fuels to biomass in order to supply citizens of Jawor with energy services they need with necessary amounts of useful energy in the right side of Tab. 5.1. What should be mentioned is that positions on the left side show primary energy potential.

There are questions that up to this point remain without answer:
- What is the most suitable conversion technology for each of the resources?
- Which resource would be most suitable to be utilised in Jawor?
- Are there any upgrades necessary for each of the fuels?

The most important connection between supply and demand is something that allows to match one to another is efficiency.

5.2 Technologies best suitable for utilisation of potential resources

Suitability and economical viability of biomass dedicated technologies are much dependant on the desired power output of the device. I think that most efficient way to provide energy services would always be poli-generation of many services. Usually it's realised by CHP (Combined Heat and Power) or CHCP (Combined Heat, Cooling and Power) plants. It also might be achieved by producing for example biofuel in combination with other chemicals and then using that fuel to produce power end heat.

Power output of the plant is an important issue, because usage of energy services changes in time and it's never constant values. It depends much on users and energy services used by them, but in general it might be said that there are always power peaks for heating and electricity. Heating has it's peak during winter when there is highest demand for that service due to highest temperature differences between desired temperature inside and temperature outside. Desired heat varies on daily basis (example in Tab 2.3) and it's necessary that system should be able to supply users with necessary amount of heat during coldest days which sets the desired power output for the devices. To calculate heating power for single household Polish Technical Standard [11] applies, although for this assessment i would only roughly estimate that demand. Standard temperatures used for calculations depend on classifying region into proper climate zone and most part of Lower Silesia, including Jawor, belongs to 1st zone (Tab 5.2).

I assume that peak power is $P_{th\,max} \cong \mathbf{38,71\ MW_{th}}$ (Appendix A)

Table 5.2 - Climate zones and standard temperatures according to Polish Technical Standards

Climate zone	Standard outside temperature °C	Mean outside temperature °C
I	−16	7,7
II	−18	7,9
III	−20	7,6
IV	−22	6,9
V	−24	5,5

Figure 5.1 - Electric power demand for Poland on yearly basis [13]

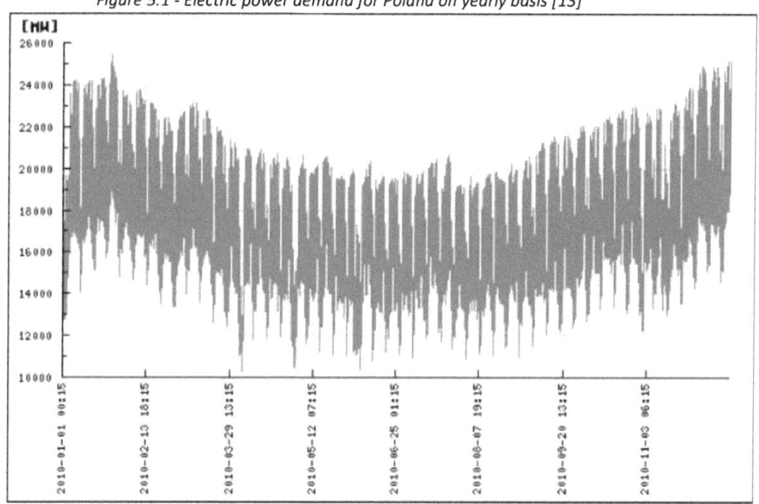

Figure 5.2 - Electricity power demand for Poland on 1st of march 2010 [13]

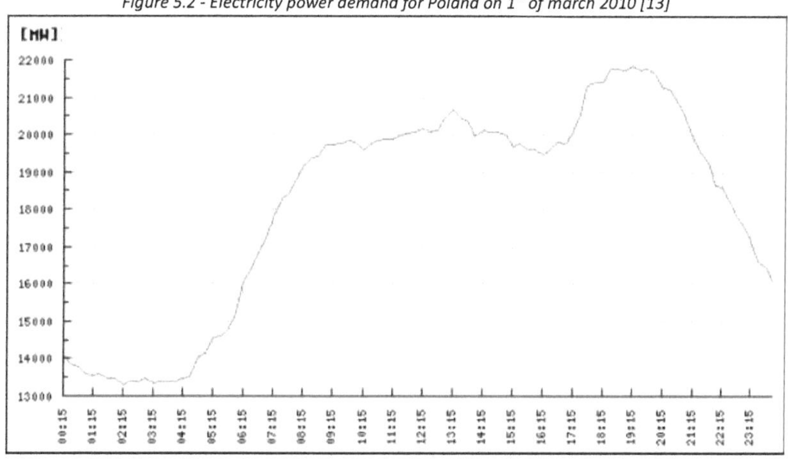

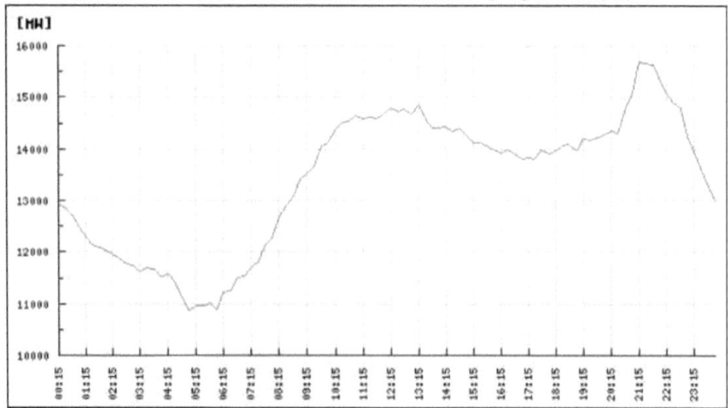

Figure 5.3 - Electricity power demand for Poland on 1ˢᵗ of august 2010 [13]

Electricity demand varies depending on time of the day (night) and time of the year. Figures 5.2 and 5.3 bring a conclusion that higher power demand during shorter day (march - fig 5.3) shows how much electricity is used to produce the light for households and offices (it might also mean high energy saving potential in terms of light source changing for energy saving ones). In my opinion low usage of electricity during longer day period in august might support that as well, also it might confirm low usage of electricity for cooling purposes cause august is usually one of the warmest months of the year.

In my opinion figures 5.1 , 5,2 and 5.3 show that assuming peak power for electricity at 150% of mean power is reasonable.

$$P_{el\ max} = \frac{Q_{el}}{t_{year}} \cdot 1{,}25 = \frac{11{,}4}{8\ 760} \cdot 1{,}25 \cong \mathbf{1{,}63\ MW_{el}}$$

5.3 Identification of most suitable resource and technology

In my opinion technology used should meet few necessary requirements. It should be commercially available and well proven, it should be environmental friendly - meeting all necessary emission requirements and it has to affordable in terms of economy.

Table 5.1 clearly shows that transfer into bioenergy is possible in Jawor. Demand for heat is large comparing to demand for electricity. The reason for that is different ways i used to estimate those types of useful energy. Estimation for electricity was made only for households and small entrepreneurs. Estimation of necessary heat was based on output from heat plant which delivers heat not only to households but also for public facilities, swimming pool and to industrial facilities of the town (although it's not a process heat but just comfort heating for those facilities).

In my opinion energy from MSW, dung and sewage sludge is the one that should be utilised in the first place because it brings other environmental benefits such as reduction of waste dumped into the landfill which is a necessity for Jawor (Tab. 4.4).

My proposition is technology that allows to produce RDF pellets and biogas like Wabio [10] or Shubio (Appendixes B and C). Pellets could be combusted in CHP plant. Biogas could be combusted in reciprocating engines to produce electricity necessary for the processes and some extra heat. According to [5] there is a possibility to mix up to 10% of biogas ($60\% CH_4, 40\% CO_2$) with natural gas in the system to supply gas for cooking. There is also a possibility to clean biogas in a scrubber and then put biomethane into the pipelines.

Figure 5.3 - Schubio process flow diagram

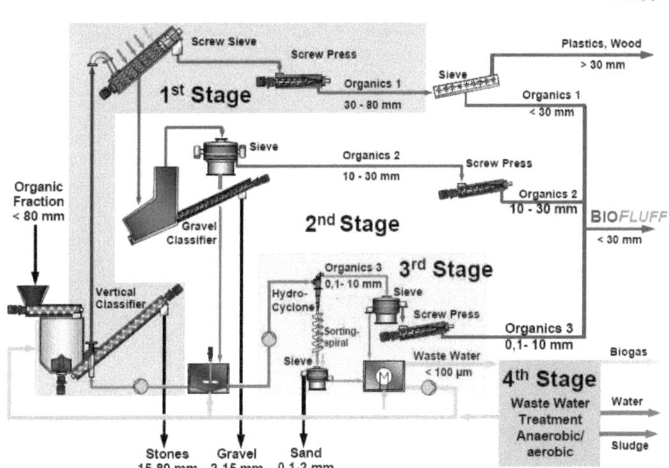

For sewage sludge and dung i would also recommend either production of biogas and it's combustion in reciprocating engine for combined power and heat production or upgrading it for biomethane.

I think that the best feedstock for the CHP plant would be **pelletised straw**. Wood resources are located in further part of the region and there is a competition because for that resource mainly because of big CHP unit, in capital of the region Wrocław, that co-combust biomass. There are forests in northwestern part of the region but in the nearest future there would be probably high competition because of condensing power plant in Turów near Bogatynia (Figure 5.4). It's being run on brown coal mined there but there are plans to co-combust biomass there as well.

Figure 5.4 - Potential and utilisation of bioenergy in Lower Silesia [4]

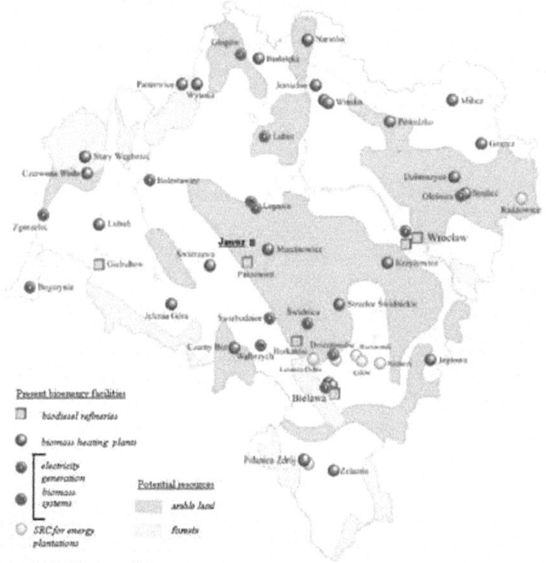

Also surrounding areas are full of arable land which would make transport cost quite low. I think that upgrade of the straw to pellets is necessary to assure proper quality of fuel since straw is a fuel that causes a lot of problems during combustion. Due to low ash melting point fluidised bed combustion technology would be suitable.

I assume (Appendix A)

$$P_{th\,max} \cong 38,71\ MW_{th}$$

and thermal efficiency of the plant 55 %

Nominal power of the device should be:

$$P_n \cong \frac{38,71\ MW_{th}}{0,55} \cong 70,38\ MW$$

My recommendation would be **CHP plant with 70 MW Circulating Fluidised Bed boiler** situated in the same location as one of the present heating plants located more to the north east of those two (Figure 5.5).

Figure 5.5 - Map of Jawor with marked locations for present and future facilities [14]

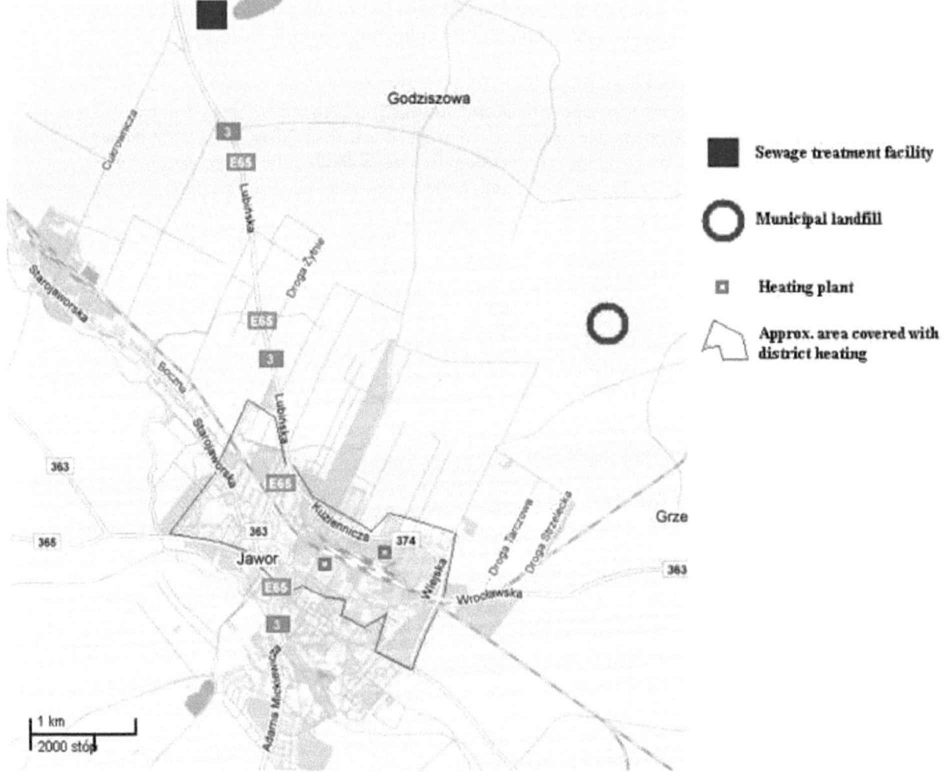

6. Conclusions and comments

It's never easy to make changes in order to replace fossil fuels with biomass. One needs to fit into local conditions and the more changes need to be made, the higher cost would be. To achieve final result i made additional assumption that rest of the town would be connected to district heating. Since Poland needs to increase bioenergy share in country's energy mix investment should pay off. Other environmental benefits like decreasing the amount of waste dumped into landfills might predestine some investment project to participate in EU environmental programmes subsidising such investments. CHP technology recommended would probably reach 30 - 35% electric efficiency. Amount of necessary heat is high comparing to the amount of electricity. Although there is some overestimation involved it also shows possibility to sell some electricity to the grid, which could be bought for example by local industry.

Since CHP unit would supply town with sufficient heat amounts, biogas from MSW, dung and sewage sludge should be utilised in other way than cogeneration. Except energy necessary for own supply surplus biogas should be upgraded to biomethane and either mixed with natural gas and put into system or compressed and used as a vehicle fuel. Retrofitting vehicles to run on CNG is well proven technology and could be done even in small workshops. Since landfill and are both far from town and from each other there would be necessity either to build special pipelines between them and natural gas system facilities. Other alternative would be to build compressing station on site. That would also give a possibility of selling that gas to the surrounding villages, because some of those terrains do not have natural gas pipeline connections, therefore it could also bring some bioenergy to villages surrounding Jawor for heating purposes.

Main conclusion is: transfer from fossil energy to bioenergy in Jawor is possible.

7. References

[1] "Plan gospodarki odpadami dla powiatu Jawor" Dorota Jeremicz for Jawor county administration; 2007 (eng. - Waste management plan for Jawor county)
[2] "Regionalny program operacyjny polityki leśnej państwa" Regionalna dyrekcja lasów państwowych; 2003 (eng. - Regional operational plan for state silvicultural policy - published by Regional Division of State Forests Agency)
[3] "Krajowy plan działania w zakresie energii ze źródeł odnawialnych" Ministry of Economy; 2010 (eng. National plan for sustainable energy sources)
[4] "Potencjał Dolnego Śląska w zakresie rozwoju alternatywnych źródeł energii" report for Regional administration; 2006 (eng. Lower Silesia potential for development of sustainable energy sources)
[5] "Analiza możliwości mieszania biogazu z gazem ziemnym z uwzględnieniem limitów wymaganej jakości gazu sieciowego" W.Kostowski, K.Górny; Instal 3/2010 (eng. Analysis of possibility of mixing biogas with natural gas with respect to limitations set by desired gas quality in system networks)
[6] http://www.stat.gov.pl/bdl/html/indeks.html - Główny Urząd Statystyczny Bank Danych Lokalnych (eng. Central Statistical Office Regional Database)
[7]http://bape.com.pl/eleff/Kalkulator.aspx - EL-EFF Region: Poprawa efektywności wykorzystania energii elektrycznej w 8 regionach Europy (eng. EL-EFF Region: Boosting efficiency o electricity use o 8 European Regions)
[8] „Krajowy program zwiększania lesistości" Polish government, 1995 (eng. National Forestation Programme)
[9] "Oszacowanie potencjału zmniejszenia zużycia energii elektrycznej w gospodarstwach domowych w Polsce" J.Wojtulewicz, A.Osicki (eng. Estimation of potential reduction of electricity usage in households in Poland).
[10] "Brilliance of bioenergy in business and in practice" R. Sims 2006
[11] "PN-EN 12831:2006. Instalacje ogrzewcze w budynkach – Metoda obliczania projektowe-go obciążenia cieplnego." - Polish Technical Standard (eng. Heating installations in buildings - method to calculate heat load in the projects)
[12] "Prognozowanie zapotrzebowania na letnią moc szczytową dla krajowego systemu elektroenergetycznego" J.Łyp, T.Popławski, K.Dąsal - Polityka energetyczna 2009 (eng. „Prognose of demand for peak electricity power in summer for national electricity-power system" - energy policy periodic)
[13] http://www.pse-operator.pl/index.php?dzid=77 - website of Polish Electricity Grid Operator
[14] http://maps.google.pl/

8. Appendix

A - peak heating power estimation

$$Q_{th} = 156{,}1 \; {}^{GWh}\!/_{year}$$

Amount of heat distributed to households with district heating network is dependent on temperature outside, but district heating facility also produces heat for hot tap water which is dependent on outside temperature only in terms of pipeline heat loses during transport (which has already been included since for estimations we used old facility output values which already included heat loss during pipeline transport since facility output needs to meet demands for energy service users - households).

That is the reason why i introduced coefficient $b = \dfrac{heat\ used\ for\ comfort\ heating}{total\ heat\ produced}$

I assume $b = 60\%$

I assume that heating season 8 months per year.

$$t_{season} = \frac{8}{12} \cdot 365 \cdot 24 = 5\ 832\ h$$

$$c = 365 \cdot 24 = 8\ 760\ h$$

$$\bar{P}_{th} = \left(\frac{b \cdot Q_{th}}{t_{season}} + \frac{(1-b) \cdot Q_{th}}{t_{year}}\right) = \left(\frac{0{,}6 \cdot 156{,}1}{5\ 832} + \frac{0{,}4 \cdot 156{,}1}{8\ 760}\right) \cong 0{,}02319\ GW = 23{,}19\ MW_{th}$$

$$\Delta Q = m \cdot c_p \cdot \Delta T$$

$$\bar{P} = \frac{\Delta Q}{\Delta t}$$

So taking into consideration desired temperatures in Polish technical standards [11]:

$$\frac{\bar{P}_{th}}{P_{th\ max}} \sim \frac{T_{in} - T_{mean}}{T_{in} - T_{out}}$$

$$P_{th\ max} = \bar{P}_{th} \cdot \frac{T_{in} - T_{out}}{T_{in} - T_{mean}} \cdot b$$

T_{in} - desired temperature inside; i assume 21 °C.
T_{out} - project temperature outside; according to [11] - 16 °C.
T_{mean} - mean temperature outside; according to [11] 7,7 °C.

$$P_{th\ max} = 23{,}19 \cdot \frac{21 - -16}{21 - 7{,}7} \cdot 0{,}6 \cong \mathbf{38{,}71\ MW_{th}}$$

Which is necessary thermal power to produce necessary heat during peak load.

B - MSW to biogas (Carbon efficiency)

I assume using process similar to Wabbio (according to [10] more than 500 kg RDF pellets could be achieved from 1 tonne of MWS).

I assume that:
50% by weight could be converted to RDF
30% by weight is the material that after sorting goes to landfill (loses)
20% by weight is bioreactor feed

Base data for biogas:
Dry substance in bioreactor feed: 10%
Higher heating value: 19,5 MJ/kg
Ash content: 50%
Digestible carbon content 50%
Raw material input is $20\% \cdot 5\,650\frac{t}{year} = 1130\ t/year$
External energy input to biogas reactors (heat + stirring energy): 120 GJ / year
Carbon loss to solids 20%
Carbon conversion 60 %

Carbon inflow:
$$10\% \cdot 50\% \cdot 50\% \cdot 1130\ t = 28,25\ t$$

Carbon left after loss to solids:
$$(100 - 20)\% \cdot 28,25\ t = 22,6\ t$$

Conversion to methane:
$$60\% \cdot 22,6\ t = 13,56\ t$$

Carbon to CO_2:
$$28,25 - 22,6 = 9,04\ t$$

MC = 12 g/mole = 12 t /Mmole

Carbon inflow: $\quad {28,25\ t}/{12\ t\ /\text{Mmole}} \cong 2\,354\ kmoles$

Carbon out to solids: $\quad {(28,25 - 22,6)\ t}/{12\ t\ /\text{Mmole}} \cong 471\ kmoles$

Conversion to methane: $\quad {13,56\ t}/{12\ t\ /\text{Mmole}} = 1\,130\ kmoles$

Carbon to CO_2: $\quad {9,04\ t}/{12\ t\ /\text{Mmole}} \cong 753\ kmoles$

****

Total carbon out: $\quad {((28,25 - 22,6) + 13,56 + 9,04)}/{12\ t\ /\text{Mmole}} \cong 471 + 1130 + 753\ kmoles$

$$\frac{28,25\ t}{12\ t\ /\text{Mmole}} \cong 2\,354\ kmoles$$

<u>Energy in:</u>

Energy from DAF substance: $10\% \cdot 50\% \cdot 1130\ t \cdot 19{,}5\ {}^{GJ}\!/_{t} = 1\ 101{,}75\ GJ$

External energy in: $120\ GJ$

Oxidation loss in CO_2: 0,393 GJ / kmole
Chemical energy in methane: 0,800 GJ / kmole

Energy out:

In methane: $1\ 130\ kmoles \cdot 0{,}8\ {}^{GJ}\!/_{kmole} = \mathbf{904\ GJ}$

Energy loss via CO_2: $753\ kmoles \cdot 0{,}393\ {}^{GJ}\!/_{kmole} \cong 256\ GJ$

There is a 61,75 GJ (approx. 5%) difference between in and out which may be accounted as a loss of heat and inaccuracies.

Total amount of gas is: $1130 + 753 = 1\ 883\ kmoles$

C - MSW to RDF

Raw material input is $50\% \cdot 5\ 650\ \dfrac{t}{year} = 2\ 825\ t/year$

I assume that for pelletising process part of pellets was consumed for energy necessary for the process - that is remaining RDF pellets are 25% of input material.

RDF pellets output:
$$2\ 825\ t \cdot 0{,}25 = 706{,}35\ t$$

I assume heating value of RDF pellets 13 GJ / $tonne$

$$706{,}35 \cdot 13 \cong \mathbf{9181{,}25\ GJ}$$

YOUR KNOWLEDGE HAS VALUE

- We will publish your bachelor's and
  master's thesis, essays and papers

- Your own eBook and book -
  sold worldwide in all relevant shops

- Earn money with each sale

Upload your text at www.GRIN.com
and publish for free